# LEGO

by Grace Hansen

Abdo Kids Jumbo is an Imprint of Abdo Kids
abdobooks.com

**abdobooks.com**

Published by Abdo Kids, a division of ABDO, P.O. Box 398166, Minneapolis, Minnesota 55439. Copyright © 2023 by Abdo Consulting Group, Inc. International copyrights reserved in all countries. No part of this book may be reproduced in any form without written permission from the publisher. Abdo Kids Jumbo™ is a trademark and logo of Abdo Kids.

Printed in China.

102022

012023

Photo Credits: Alamy, Everette Collection, Getty Images, Shutterstock, ©David Mellis p.7/ CC BY 2.0

Production Contributors: Teddy Borth, Jennie Forsberg, Grace Hansen
Design Contributors: Candice Keimig, Pakou Moua

Library of Congress Control Number: 2022937184

Publisher's Cataloging-in-Publication Data

Names: Hansen, Grace, author.

Title: LEGO / by Grace Hansen

Description: Minneapolis, Minnesota : Abdo Kids, 2023 | Series: Toy mania! | Includes online resources and index.

Identifiers: ISBN 9781098264284 (lib. bdg.) | ISBN 9781098264840 (ebook) | ISBN 9781098265120 (Read-to-Me ebook)

Subjects: LCSH: LEGO toys--Juvenile literature. | Building blocks (Toys)--Juvenile literature. | Toys--Juvenile literature. | LEGO koncernen (Denmark)--Juvenile literature.

Classification: DDC 688.725--dc23

# Table of Contents

| | |
|---|---|
| LEGO . . . . . . . . . . . . . . . . . . . . . . . . 4 | Glossary . . . . . . . . . . . . . . . . . . . . . 23 |
| Play Well . . . . . . . . . . . . . . . . . . . . . 6 | Index . . . . . . . . . . . . . . . . . . . . . . . 24 |
| Facts and Minifigures . . . . . . . . 18 | Abdo Kids Code. . . . . . . . . . . . . . 24 |
| More Facts . . . . . . . . . . . . . . . . . 22 | |

## LEGO

Piece by piece, LEGO bricks stack up to create more than cool displays. LEGOs help kids build confidence and imagination.

## Play Well

In 1934, Ole Christiansen was hard at work in the tiny town of Billund, Denmark. Ole was a carpenter. Everyone liked his wooden household goods and furniture. But parents in Billund loved Ole's durable kids' toys.

Legoland in Billund, Denmark

Ole wanted a company that focused on toys. But he needed a name. The Danish phrase *leg godt* means "play well." Ole combined the words to make the name LEGO.

For many years, all LEGO toys were carved from wood. In the mid-1940s, plastic became a new material for toy makers. In 1947, Ole bought a machine that could make lots of plastic toys fast!

The LEGO team designed colorful plastic building blocks. The blocks could be put together and pulled apart. In 1949, the company began making Automatic Binding Bricks.

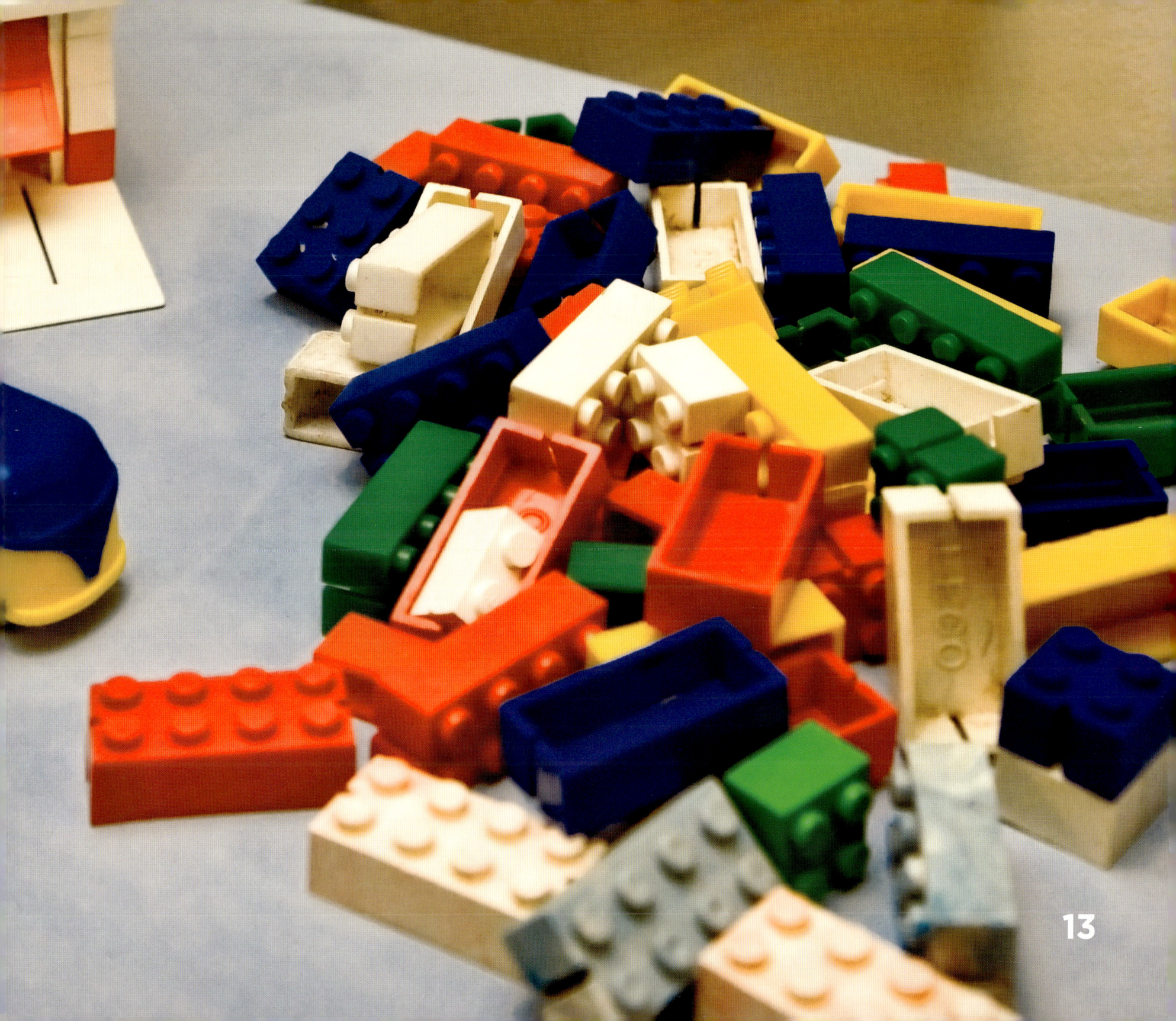

In 1955, LEGO launched its first System of Play. It was called Town Plan No. 1. The system was very popular. But sometimes the pieces didn't stick together very well.

On January 28, 1958, the new and improved LEGO Brick was **patented**. The brick had a stud-and-tube locking system. Building possibilities became endless.

## Facts and Minifigures

LEGO minifigures were introduced in 1970. Today, there are more than 8,000 different characters to choose from! The beloved minifigures have even starred on the big screen.

LEGO has hundreds of play sets that span many themes. From **skyscrapers** to superheroes, LEGO has it all. But LEGO makes it possible for people to build whatever they can dream of!

## More Facts

- In 2012, Guinness World Records named LEGO the largest tire manufacturer in the world. Each year, 306 million tiny rubber tires are made for LEGO!

- The first Legoland opened in June 1968 in Billund, Denmark. Today, there are Legoland parks around the world in places like Japan, the United States, and the United Kingdom.

- LEGO sells a lot of bricks each year. If you placed them end-to-end, they would circle the Earth more than five times!

# Glossary

**carpenter** – a person who builds or repairs wooden structures or their parts.

**confidence** – a sense of trust or faith in oneself.

**durable** – not easily broken or worn out.

**imagination** – the act or power of the mind to form a thought, picture, or image of something.

**patented** – received a government grant that gives one the right to make, use, or sell an invention.

**skyscraper** – a very tall building.

# Index

architecture 20

Billund, Denmark 6

bricks 4, 12, 14, 16, 20

Christiansen, Ole 6, 8, 10

minifigures 18

movies 18

name 8

plastic 10

play sets 14, 20

skyscrapers 20

superheroes 20

wood 6, 10

Visit **abdokids.com** to access crafts, games, videos, and more!

Use Abdo Kids code **TLK4284** or scan this QR code!